CREEK CRITTERS

by Jennifer Keats Curtis with Stroud Water Research Center

illustrated by Phyllis Saroff

"Lucas, let me show you how bugs can tell a story of clean water," I say to my little brother, wiggling my eyebrows. He hates that.

Lucas rolls his eyes but follows me to the house. We grab the same tools that scientists use: rubber boots, a net, a bucket, and small paintbrushes.

We tug on our boots and head back to the freshwater creek.

Galumph! We step right into the water. This part of the creek is called a riffle. It's shallow and the water runs fast enough over the rocks to make a bubbling noise. Downstream is a pool. It's deep and the water is calm.

I splash water into our bucket and plop it next to a tall sycamore tree. Quickly, I spin around before Lucas can push me in. That water is cold!

Like a team of scientists, we work together, searching for bugs that can only live in healthy water. Aquatic macroinvertebrates (or macros for short) is the fancy name for these creatures. If you break down the word, it's easy to understand. Aquatic means water. Macro means big; in this case, big enough to see with our eyes. No microscope needed. Invertebrate means no backbone.

Young macros, called larvae or nymphs, live in water. As they grow older, some kinds of macros change—they go through metamorphosis. These adults fly around and live near the water, while other adults may spend their whole lives underwater.

I yank up a rock. Lucas and I check the slimy bottom. Nothing. We wade over two steps. I snag a smaller grey stone for inspection.

"There's some weird-looking crawler with big eyes on there," Lucas says very unscientifically.

I carefully scrape the bug into the bucket. He's a small but mighty dragonfly nymph! The young dragonfly spurts around as if he's wearing a jet pack. He sucks water in and spurts it out back, zooming from one side to the other. We put some sticks and leaves in the bucket so he can hide.

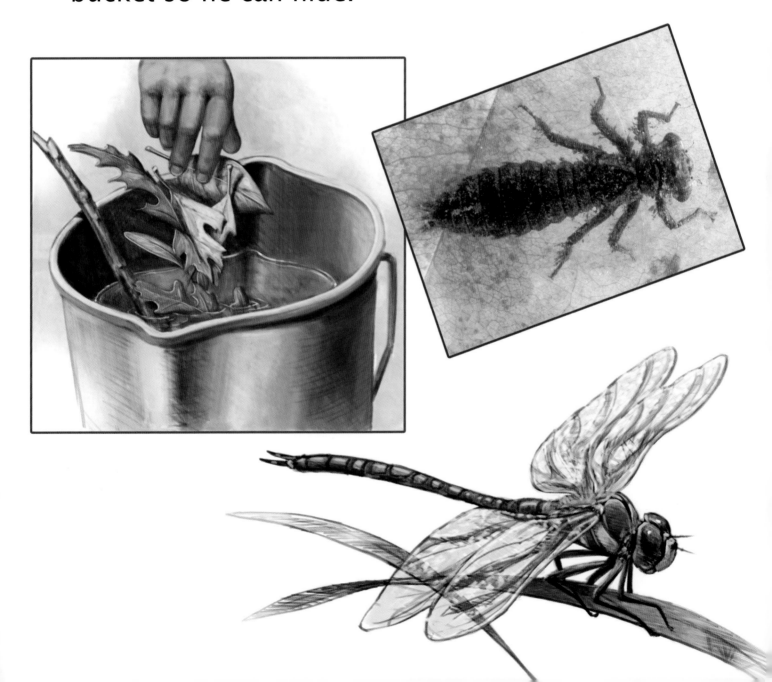

Lucas flips over another rock. I spy what looks like the smallest turtle I've ever seen. It's hanging on for dear life. It's actually a water penny, a beetle larva whose back looks like a turtle's shell. These beetles don't like creeks with pollution or muddy buildup (sediment). We put the water penny in the bucket.

You can probably tell; these camouflaged creek critters are hard to spot. They can be as small as your eyelash or as big as your thumb. They hide under, in, and around stones and pebbles. Some skitter into large bunches of leaves, called leaf packs. You can scoop up the pack with the net and slowly sift through it to check what's in there. We really want to find mayflies, stoneflies, and caddisflies. These bugs are the most sensitive to pollution. They often cannot live in dirty water. If we find them, we'll know our creek is healthy.

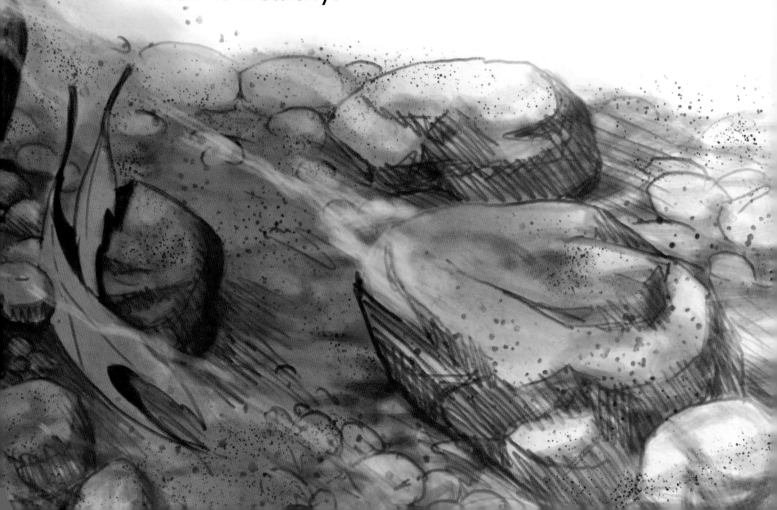

Lucas and I push the net underwater along the creek's edge. We shuffle our feet, kicking around so lots of stuff floats into our net. I pull up a hunk of rotting leaves, one of the best bug-hiding places.

Of course, Lucas grabs a slimy leaf and flings it at me. Ugh, Lucas is always doing stuff like that. I wipe off the brown goo with a groan.

We quietly stare into the net. It's hard, but you have to be patient. The itty-bitty bugs are almost the same color as the leaves and are so small that you have to look for movement to find them.

There! A caddisfly larva pokes his head out of a homemade case of rocks, leaves, and twigs. This crafty critter creates his house using sticky silk as underwater glue. Lucas gently plunks him into the bucket. Before moving on, we examine the rest of the net to make sure there are no more bugs. They'll die if they dry out.

Now it's Lucas's turn to scoop with the net. We shuffle our feet again and a heap of grey, gooey leaves glides in. As we pluck out the last leaf,

Lucas spots a six-legged bug with three tails and fluttering gills. It's a mayfly nymph! We add him to the bucket where the other bugs creep, shuffle, curl up, and hide.

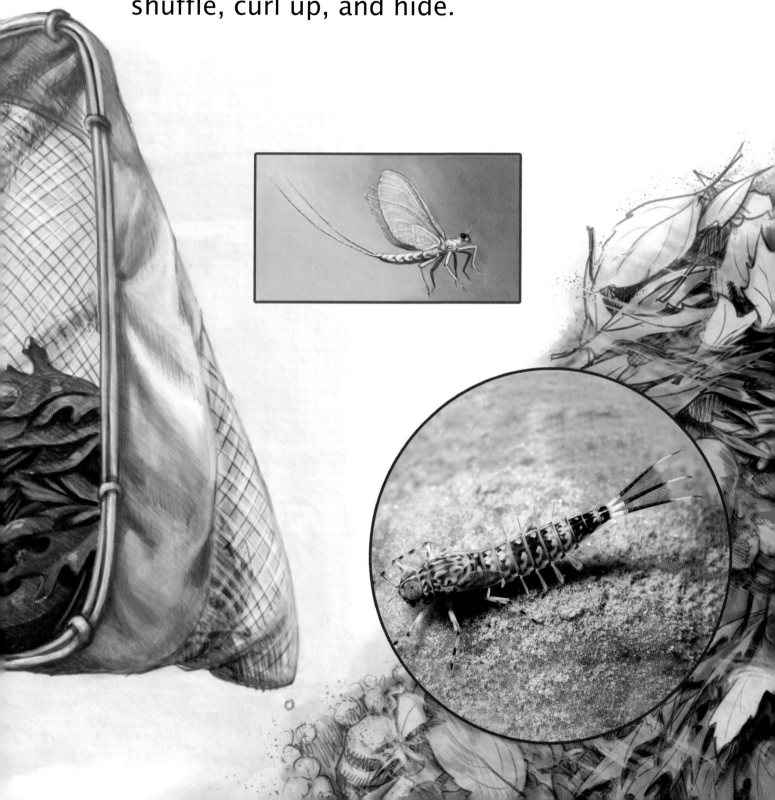

I peer into the bucket but don't see the dragonfly. There he is, stuck to a stick! We can't leave him in there too long. Dragonflies are predators; they may eat our other new friends.

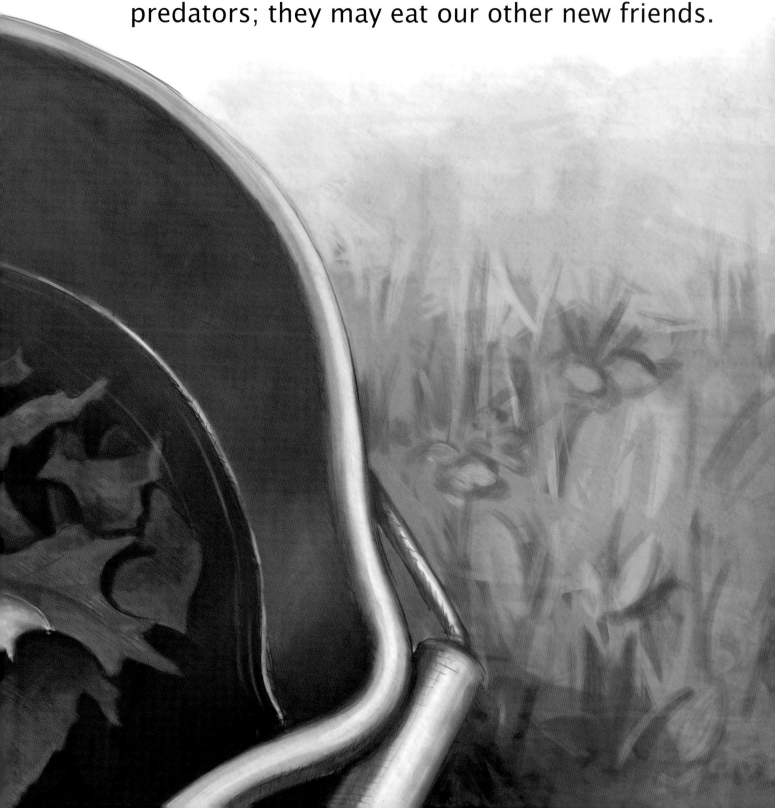

Although I really want today's adventure to include a stonefly nymph, we have found special macros, such as the mayfly and caddisfly, that live in clean water. Lucas now knows how water bugs can tell tales.

With that in mind, my brother and I decide to call it a day. We take our story-telling macroinvertebrates back down to the creek.

I smile and, together, we turn the bucket over into the water, returning the bugs to their home.

For Creative Minds

This section may be photocopied or printed from our website by the owner of this book for educational, non-commercial use. Cross-curricular teaching activities for use at home or in the classroom, interactive quizzes, and more are available online.

Visit **www.ArbordalePublishing.com** to explore additional resources.

Scavenger Hunt: Identify the Bugs

| water penny larva | stonefly nymph | dragonfly nymph |

The answers are on the copyright page.

Matching Young to Adults

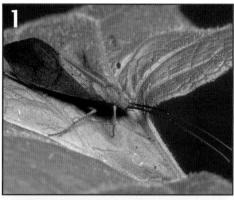

Can you match the young water bugs, called larvae or nymphs, to their adults?

Aquatic macroinvertebrates are "bugs" that spend some or all of their lives underwater, are big enough to see with the naked eye, and don't have backbones.

These four kinds of macroinvertebrates are all insects that begin their lives underwater. After the larvae or nymphs change through metamorphosis, the adults fly around and live near the water.

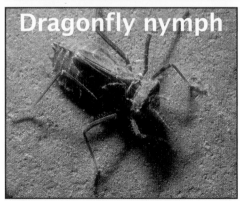

Answers: mayfly-4, caddisfly-1, stonefly-2, dragonfly-3

Scientist's Field Notebook

Explore a small section of your stream to see if it is a good place for creek critters to live! For each section below, check off the boxes of what you observe and find. Have fun!

Don't forget to bring: copy or download (www.arbordalepublishing.com) this page to take with you, pencil, clear plastic cup (or mason jar) to take a sample of stream water, shoes for getting in the stream and exploring the nearby forest, and a buddy!

Your Name: _____ *Date:* _____

Stream/River Name: _____ *Time:* _____

Current Weather
- ☐ Sunny
- ☐ Cloudy
- ☐ Partly Cloudy
- ☐ Rain
- ☐ No Rain

Check the Stream Water Part 1

Water Color
- ☐ Clear
- ☐ Brown
- ☐ Green
- ☐ Orange
- ☐ Blue
- ☐ Other: _____

Smell
- ☐ Dead fish
- ☐ Rotten Eggs
- ☐ Chlorine
- ☐ Gas/Oil
- ☐ Nothing
- ☐ Other: _____

Clarity
- ☐ Clear
- ☐ Cloudy

Fill up a clear cup with stream water for these tests!

I Also See:
- ☐ Fish
- ☐ Plastic
- ☐ Bridge
- ☐ Farm Animals
- ☐ Tires
- ☐ Buildings
- ☐ Birds
- ☐ Road
- ☐ Other: _____

Check the Stream Water Part 2

Water Surface
- ☐ Nothing
- ☐ Algae
- ☐ Oil
- ☐ Foam
- ☐ Other: _____

Water Level
- ☐ Flood
- ☐ High
- ☐ Normal
- ☐ Low
- ☐ Dry

Site Notes:

Is Your Creek Healthy? Ask the Critters!

Aquatic macroinvertebrates (macros) can be grouped by how sensitive they are to pollution or dirty water. A healthy creek should support many sensitive macros. Macros in every group have special jobs in the creek. Scientists want to find many different kinds of macros and lots of each kind!

Group 1: Sensitive — These aquatic macroinvertebrates need clean water to survive.

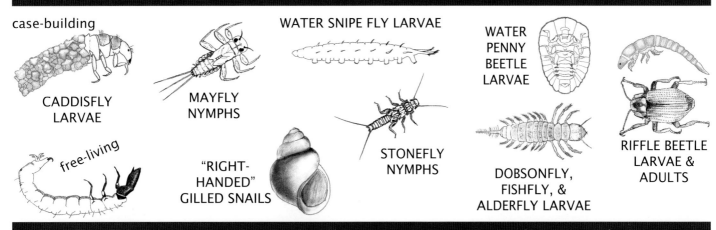

Group 2: Somewhat Sensitive — These aquatic macroinvertebrates can live in somewhat polluted water.

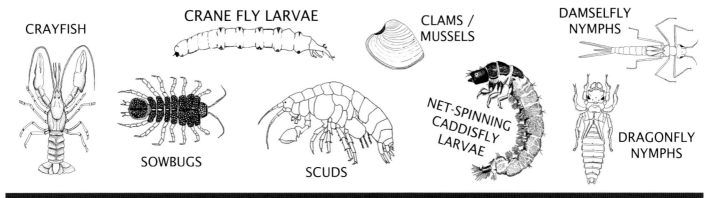

Group 3: Tolerant — These aquatic macroinvertebrates can live in very clean, somewhat clean, OR polluted water!

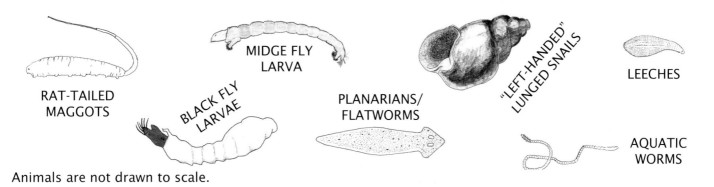

Animals are not drawn to scale.

Thank you to Steve Kerlin, Mandy Nix, and Tara Muenz of Stroud Water Research Center as partners in the creation of this book and for verifying the scientific accuracy.

All photographs were provided by Stroud Water Research Center.

Bibliography:

Macroinvertebrates. Learning to See, Seeing to Learn. Macroinvertebrates.org. Carnegie Mellon University. Website. February 2, 2019.
Our Focus Is Fresh Water. Stroud Water Research Center. Website. April 3, 2017.
Watershed Education: Interpreting Biological Results. Pennsylvania Department of Conservation and Natural Resources. Website. January 30, 2019.
Welcome to the Leaf Packet Network. Stroud Water Research Center. 2019. Website. October 16, 2018.
White Clay Creek Wild & Scenic River. White Clay Wild & Scenic River Program/White Clay Watershed Association, Environmental and Cultural Resources. Website. February 1, 2019.
WikiWatershed: Web Tools Advancing Knowledge and Stewardship of Fresh Water. Stroud Water Research Center. December 18, 2018, Website. January 30, 2019.

Library of Congress Cataloging-in-Publication Data

Names: Curtis, Jennifer Keats, author. | Saroff, Phyllis V., illustrator.
Title: Creek critters / by Jennifer Keats Curtis with Stroud Water Research Center ; illustrated by Phyllis Saroff.
Description: Mt. Pleasant : Arbordale Publishing, LLC, [2020] | Includes bibliographical references.
Identifiers: LCCN 2019027428 (print) | LCCN 2019027429 (ebook) | ISBN 9781643517483 (hardcover) | ISBN 9781643517537 (trade paperback) | ISBN 9781643517834 (English interactive ebook) | ISBN 9781643517735 (epub)
Subjects: LCSH: Stream ecology--Juvenile literature.
Classification: LCC QH541.5.S7 C867 2020 (print) | LCC QH541.5.S7 (ebook) | DDC 577.6/4--dc23
LC record available at https://lccn.loc.gov/2019027428
LC ebook record available at https://lccn.loc.gov/2019027429

Lexile® Level: 660L

key phrases: Environmental Education, water quality, Ciitzen Science

Title in Spanish: **Animalitos de arroyo**

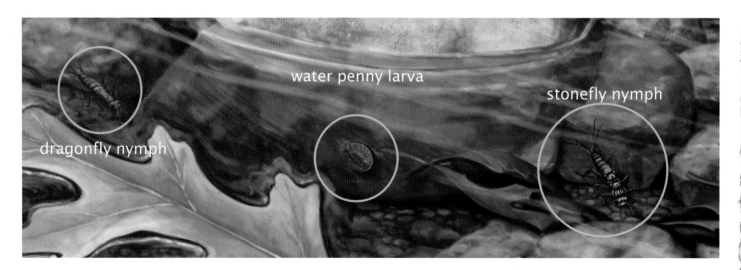

Text Copyright 2020 © by Jennifer Keats Curtis and Stroud Water Research Center
Illustration Copyright 2020 © by Phyllis Saroff

The "For Creative Minds" educational section may be copied by the owner for personal use or by educators using copies in classroom settings.

Printed in China, November 2019
This product conforms to CPSIA 2008
First Printing

Arbordale Publishing
Mt. Pleasant, SC 29464
www.ArbordalePublishing.com